Estimation
INVESTIGATIONS

More than 65 Activities to Build Mathematical Reasoning and Number Sense

By Marcia Miller and Martin Lee

SCHOLASTIC
PROFESSIONAL BOOKS

New York • Toronto • London • Auckland • Sydney

For Hannah and Clark

Cover design by Vincent Ceci

Cover photograph by Donnelly Marks

Interior design and illustrations by Drew Hires

ISBN 0-590-49602-6

12 11 10 9 8 7 6 5 4 3 2 1   4 5 / 9

Printed in the U.S.A.

Contents

Introduction

Children and adults estimate constantly. We rarely make it through a day without guessing, approximating, or predicting answers to questions like these:

◆ *How much money should I bring when I go to the store?*

◆ *When should I leave so I get to the meeting on time?*

◆ *How many students are needed to carry the equipment to the field?*

◆ *Which is the fastest route home during rush hour?*

◆ *When should I start the activity so that students finish before lunch?*

◆ *How much wrapping paper do I need?*

◆ *Which container will hold all the leftover pasta?*

◆ *How much chicken should I buy?*

Estimation is a part of our everyday life *and* a powerful mathematical idea. We estimate to solve problems. We estimate because we have to make so many of our daily calculations without pencil and paper or calculators. Although quick answers are just a few keystrokes away when we do use a calculator, so are staggering errors! Then we can estimate to spot unreasonable answers. But perhaps most importantly, we often estimate because approximations can make better sense than exact answers.

Estimation belongs in every classroom. The National Council of Teachers of Mathematics (NCTM) thinks so, too. In the *Curriculum and Evaluation Standards for School Mathematics*, both Standard 5 for grades K–4 and Standard 7 for grades 5–8 stress the importance of including estimation in the curriculum. According to the NCTM Standards, estimation should be an "...ongoing part of [the] study of numbers, computation, and measurement." NCTM goes on to say that students should understand what is meant by an estimate, when it makes sense to estimate, and how precise an estimate is needed in a given situation.

NCTM also asserts that "Instruction should emphasize the development of an estimation mind-set." This mind-set is well worth developing. Children who appreciate that math is more than finding exact answers can be more flexible and creative when working with numbers. They can gain a feeling of power over numbers, which can lead to new insights into concepts and procedures and to a more confident attitude toward dealing with everyday mathematical situations.

Let's help children become better estimators—for today and for the future.

About This Book

Most of today's mathematic texts provide instruction in computational estimation strategies. The focus of *Estimation Investigations* is different. Our goal is to provide open-ended questions that challenge students to develop that estimation mind-set. Some investigations pose common problems anyone might face. Others are non-routine, even whimsical. Some are brief and can be solved using number sense, spatial sense, or just plain common sense. Others will take a little longer and require trying more than one approach. And many are Fermi problems, named for the great physicist, who enjoyed posing problems that would seem impossible to solve at first glance, but could become manageable when broken down into smaller, discrete parts.

As students work through these investigations, they will encounter problems with varying degrees of difficulty that span a range of mathematical topics. Students will be called upon to use all of the computational methods they know—mental math, calculators, paper and pencil, or a combination of them. Regardless of the length, complexity, or particular mathematical strands involved, all the investigations in the book encourage children to be creative, flexible, and holistic thinkers. All the questions posed can be solved; yet few have simple, "right" answers.

Scan the contents of *Estimation Investigations* and flip through the pages. You'll notice that the book is divided into four parts. You'll see a variety of investigations grouped according to how long we estimate each is likely to take. Note that we have listed broad mathematical connections for each investigation in Parts 2 and 3.

You'll notice that a ☐ starts the teacher notes for each investigation. Use it to mark off problems and activities as you use them.

You'll find a ✔ with each investigation throughout the book. This check mark signals a suggested reasonable estimate, provided whenever possible.

Part 1: Take Five poses brief estimation questions to ask students. Although some may merit longer investigation and may need to be completed outside of class, all are designed to take no longer than 10 minutes of class time. Use the Teaching Tips provided.

Part 2: Take Twenty contains reproducible activities estimated to take about 15–25 minutes of class time. Duplicate and distribute them to individuals, pairs, or small groups. Like the Take Five activities, each of these is supported by Teaching Tips.

Part 3: Take Longer has reproducible estimation activities that might take a whole math class or longer to complete. They are designed for pairs or small groups, but could be done by individuals if you wish to assign them that way. All contain Teaching Tips.

Part 4: Appendix is a concise resource that presents computational estimation strategies, with brief explanations and examples with annotated solutions. Consult this section for a quick reminder of five commonly taught techniques. You may guide students to use any of these that make sense as they work through the investigations in this book.

Teaching Tips

Move about the book as you see fit, presenting the activities in any order that makes sense for your class or schedule.

◆ You may find that some questions are too advanced for your class, while others may be more basic. Some may take more time than you have. Some will take less time than you planned. So, revise them to suit your students, or extend them if results suggest further investigation. Look to the Teaching Tips for ways to alter or extend the investigation.

◆ Use the **Take Five** investigations as warm-ups, problems of the day, or assign them for homework. Use them as a way for students to get acquainted with group work.

◆ Invite students to work on any of the investigations at home with family members.

◆ Allow time for groups to present their estimates and describe their strategies to the class, particularly in the longer open-ended activities. Always encourage talking, questioning, sharing, writing, and summarizing.

◆ Have grid paper, rulers, tape measures, containers, counters, scales, cups, string, clocks or stop watches, and any other useful materials available for students.

◆ Set up an Estimation Corner in your classroom. Compile investigation kits, each containing a laminated estimation investigation and any tools or materials needed. Keep them in the Corner for students to use.

◆ Make a file box to hold estimation investigations you write on index cards. Keep it in the Estimation Corner. Add cards any time. Invite students to add ideas of their own.

◆ Use the ideas in the **Appendix** to develop activities or mini-lessons to augment similar activities presented in your math textbook series.

◆ Take advantage of any estimation opportunities that present themselves during the course of any typical day. Talk about them and work them out together.

◆ Some students are uneasy at first with the idea that some problems may have no exact, correct answers. Help them to become more open-minded and flexible thinkers.

◆ Be a role model—show that you, too, are an investigator who must figure out logical and effective ways to estimate.

◆ Above all, focus on that estimation mind-set. Encourage students to be inventive in their approaches to the problems, to take risks, and to try different techniques. Guide them to evaluate their estimates as they would any solutions and to make adjustments as necessary. Support their creative efforts and applaud unique approaches.

Now, take a deep breath and leap into some *Estimation Investigations*!

PART 1
Take Five

1. How long does it take to make a peanut butter and jelly sandwich?

Teaching Tips:

◆ Ask students to assume that they have all necessary ingredients and utensils at hand. The estimate should account for preliminary steps—collecting ingredients and utensils, as well as cleaning up and putting everything away afterwards.

◆ Pose a similar question for other simple foods students might prepare, such as scrambling eggs or creating an ice cream sundae.

 3–10 minutes makes sense.

2. What weighs more—your shoe or your math book?

Teaching Tips:

◆ After students estimate, they can check their responses on a scale.

◆ Compare pairs of other familiar objects to help students develop a greater sense of relative weight. For instance, how about a shoe vs. a stapler, an orange vs. a chalkboard eraser, or an umbrella vs. a box of crayons?

 *Accept any responses students can justify.*

3. How far is 10 feet?

Teaching Tips:

◆ Ask students to show their estimates by placing markers on the ground. Measure to check their estimates. Repeat several times. Do estimates improve? You can present this as a game, with the winner being the student or group with the closest estimate.

◆ Try with other lengths, such as 1 yard, 25 feet, 10 meters, etc. Vary the setting by trying this investigation in other parts of the classroom, in the gym, in a corridor, or outside.

Most students improve their measurement sense the more they try estimates like this.

4. How long will it take you to do all of tonight's homework?

Teaching Tips:

◆ Discuss reasonable intervals for the estimate, such as the nearest quarter-hour or the nearest 5 minutes.

◆ Students may work in pairs to discuss how to estimate. After doing so, each student can make his or her estimate, record it, and leave it in school. That evening, students should record the actual time spent doing their homework and make comparisons the following day.

◆ Encourage students to predict and check the duration of other daily activities.

✔ *Accept any reasonable estimate students can justify.*

5. How many books are in our classroom?

Teaching Tips:

◆ Encourage students to consider not only the books they can see on classroom shelves, but also textbooks or library books students may have in their desks, book bags, or lockers. They should also consider books stored in classroom closets.

◆ Students can move about the room to help them estimate, or challenge them to estimate from their seats! Emphasize that students should not actually count all the books.

◆ Students may need to use paper and pencil to record partial estimates.

✔ *Accept all estimates students can justify.*

6. Using your normal walking pace, how many steps do you take to cross the widest part of the room?

Teaching Tips:

◆ Ask students to guess, then test. Discuss reasons for variations among the estimates and the actual numbers of steps.

◆ Estimate the number of steps it takes to walk the perimeter or a diagonal of the classroom.

✔ *Estimates will vary.*

7. How many times in a day do you check the time on a watch or clock?

Teaching Tips:

◆ Everyone has a different sense of time and different reasons to know what time it is. Discuss these differences as a whole class. Assume a wide range among estimates.

◆ Discuss ways to collect the data needed to verify this kind of estimate.

◆ Try similar estimates, such as the number of times in a day (or in an hour, in a class period, in ten minutes) you stretch, yawn, swallow, blink, say the word *like,* etc.

✔ *Estimates will vary.*

8. In an hour of TV, how much time is devoted to the program itself?

Teaching Tips:

◆ Discuss the other elements included in television programming, such as commercials, station identification, public service announcements, and opening and closing credits.

◆ Have students investigate this question at home to check their estimate.

◆ Will a good estimate hold true for any hour? For any channel? Discussing these issues helps students improve their estimation skills as well as their critical viewing skills.

✔ *Prime-time shows have about 47 minutes of program per 60 minutes. The ratio is different at other times of the day.*

9. How many times is a traffic light red in a day?

Teaching Tips:

◆ Discuss useful strategies, such as observing a traffic light for a short time and multiplying to account for the entire day.

◆ Talk about ways to verify estimates, such as contacting the local police or department of highways in your area.

✔ *Estimates will vary.*

10. How long can you hold your breath?

Teaching Tips:

◆ Be alert to safety. Point out that different people have different lung capacities; also, some can draw larger breaths than others.

◆ Vary the conditions for new estimation situations. For instance, may students exhale? Should they begin with a big gulp of air or without warning? Do they stand, sit, or lie down? Does this matter?

✔ *Estimates will vary.*

11. On average, hair grows a 1/2-inch a month. How long will it take you to grow waist-length hair?

Teaching Tips:

◆ Suggest that students use the length of hair at the back of the head as the current length, and assume the average growth rate, although this varies among individuals. Some students may account for changes in height over the duration of the estimate.

◆ Finger nails grow about 0.02 inches per week. Use this fact to pose a similar question students can explore.

✔ *Accept all estimates students can justify.*

12. How long does it take to buy a tomato?

Teaching Tips:

◆ Challenge students to account for the entire process: the time it takes to get to the store, to find the produce section, to select a good tomato, to get it weighed, to wait on the checkout line, to pay, to get change, and to return home.

◆ Have students verify their estimates with a family member on a real shopping trip.

✔ *Accept any estimate students can justify.*

13. How many pennies can you hold in one hand?

Teaching Tips:

◆ Students can first estimate whether they think they can hold more or less than $1.00. If real pennies are unavailable, they can use counters or play coins to test their estimates.

◆ Discuss what a *handful* means: Do you grab pennies or pile them onto the hand? Are the pennies loose or rolled? Is the hand cupped or flat?

◆ Pose similar questions using other coins. Challenge students to determine the value of the greatest handful of each kind of coin they can hold.

 Estimates will vary.

14. How many pennies will cover a dollar bill?

Teaching Tips:

◆ Be sure students estimate *first.* If possible, provide real money students can use to test their estimates, or use play money that is life-size.

◆ Try similar investigations with other coins and with other surfaces to cover.

✔ *24 pennies (no coins extend over the edges of the bill).*

15. Which is worth more—a 1-inch stack of quarters or a 1-foot row of dimes?

Teaching Tips:

◆ Point out the distinction between a *stack* (coins in a pile or tower) and a *row* (coins lined up edge-to-edge on a flat surface).

◆ Encourage students to explain how they made their decision.

◆ Pose other stack vs. row comparisons.

✔ *A 1-inch stack of quarters is worth about $3.75; a 1-foot row of dimes, about $1.70.*

16. How many ladybugs can fit on a leaf?

Teaching Tips:

◆ Prepare a tracing of the outline of a leaf, or draw the outline of a generic leaf shape. Duplicate it so that each student or group can have a copy.

◆ Pose other surface-covering investigations, such as: How many paper clips cover a math book? How many footprints cover the classroom rug? How many grains of rice cover a base ten 100 (flat) block?

✔ *Accept any estimate students can justify.*

17. How many blades of grass are in a square inch?

Teaching Tips:

◆ Be sure students know what a square inch is. They might draw or cut out a square inch from grid paper to get a better sense of the area to consider. Students can go to a park or playing field to take a closer look at a square inch of lawn to check their estimates.

◆ Ask students to estimate the number of blades of grass in a square foot.

✔ *Estimates will vary.*

18. What fraction of our classroom floor is covered by desks?

Teaching Tips:

◆ You can adapt this question to reflect the use of percentages.

◆ Vary the investigation by having different groups consider the part of the floor covered by desks, chairs, tables, empty space, shelving, rugs, etc.

◆ Challenge students to explore similar questions about rooms and furniture at home.

✔ *Accept any estimates students can support and justify.*

19. How many gulps (swallows) will it take you to drink a glass of water?

Teaching Tips:

◆ Students should agree on the size of glass to consider and the style of drinking—chug-a-lugs, dainty sips, etc.

◆ Invite students to complete this investigation at home. Discuss the results together the next day.

◆ Pose similar questions, such as: How many bites in a burger? How many chews in a piece of taffy? How many licks in a lollipop?

✔ *Estimates will vary.*

20. How many drops of water can a spoon hold before the water spills over the edge?

Teaching Tips:

◆ Students should agree on the size and composition of the spoon (metal tablespoon? teaspoon? plastic spoon? wooden cooking spoon? baby's spoon?). They should find a way to count the drops and keep the spoon steady enough to get the best verification possible.

◆ Students can estimate, then complete this investigation in class or at home.

✔ *Estimates will vary, based on the chosen parameters and the steadiness of the spoon.*

21. How much water do you use to brush your teeth if you leave the faucet on from start to finish?

Teaching Tips:

◆ Invite students to estimate in class, then complete this investigation at home.

◆ Discuss ways to measure the amount of water used without actually running the water the whole time. Have students consider variables such as how fast to run the water, when to start and stop it, how long to brush, etc.

✔ *A range of 2–6 gallons makes sense.*

22. How many kernels are there in a cup of unpopped popcorn?

Teaching Tips:

◆ Provide a sealed bag of popcorn and 1-cup measures students can examine to help them make their estimates. Invite students to share the various strategies they use.

◆ Pose similar capacity questions, such as: How many kernels in a cup of *popped* popcorn? How many raisins in a cup? How many olives in a cup?

✔ *1300–1800 kernels is a sensible range for estimates.*

23. How long does it take to toast a slice of bread?

Teaching Tips:

◆ This investigation assumes that most students have made toast and may have some sense of how long it takes. If not, substitute another common cooking activity, such as boiling water in a tea kettle, melting butter, or making a fried egg.

◆ Explore variables that affect the estimate and the actual time, such as type of bread, kind of toaster used, degree of darkness desired, and so on.

✔ *Accept all estimates students can justify.*

24. How many peanuts (in the shell) are in a pound?

Teaching Tips:

◆ You may want to display a 1-pound bag of peanuts that groups can look at, but do NOT allow them to count the peanuts. After estimates have been made, students can verify them by counting the peanuts, maybe eating some, too!

◆ Pose similar questions about 1-pound quantities, such as: How many peanuts in a pound of *shelled* peanuts? How many grapes in a pound? How many string beans in a pound? How many dry garbanzo beans in a pound?

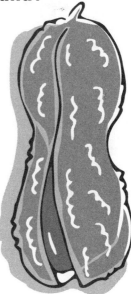

✔ *Estimates will vary depending on the size of the peanuts.*

25. Draw an irregular figure. Ask a classmate to shade 1/2.

Teaching Tips:

◆ Provide students with unlined drawing paper. The figures may be polygons or curved figures, as long as they are closed.

◆ Challenge students to repeat the investigation, shading other fractional parts, such as 1/3, 1/4, 1/10, 3/4, 2/3, or 7/8.

✔ *Have students explain how they determined what part of the figure to shade.*

26. Draw a closed figure. Shade 30%.

Teaching Tips:

◆ Provide unlined paper for students to use to draw the closed figures.

◆ Repeat, using other percentages, such as 50%, 25%, 75%, 33 1/3%, 67%, 90%, and so on.

◆ See *Shading Shapes* and *In the Shade*, pages 48–50.

✔ *Accept any shadings students can justify.*

27. If you could hike 2.5 miles an hour, how far could you hike during daylight today?

Teaching Tips:

◆ Provide sunrise and sunset times (check the daily newspaper), or guide students to find this information on their own.

◆ Students should account for time to eat and rest, as well as consider other variables to help them make or adjust their estimates. Discuss estimates together.

✔ *Estimates will vary.*

PART 2
Take Twenty

☐ How Many in a Jar?

Connections: *number sense; visual/spatial reasoning; proportional reasoning*

Teaching Tips:

◆ Collect a variety of transparent containers of different shapes (cylinders, rectangular prisms, cubes). Fill a container with objects, such as dried beans, centimeter cubes, buttons, beads, cotton balls, candies, and so on. Repeat this investigation, varying the container and filler so that students can have lots of practice sharpening their estimation skills.

◆ Talk about the strategies suggested in this book and any others students may devise. Ask them to compare the strategies and explain which ones they found most effective.

◆ Hold a weekly Container Competition. Display a filled container for a short time (say, 5 minutes) so groups can make estimates. The group whose estimate is the closest can prepare next week's container.

 *Estimates will vary.*

☐ Dot, Dot, Dot...

Connections: *number sense; visual/spatial reasoning; proportional reasoning*

Teaching Tips:

◆ Before students try this investigation, have them try a hands-on version first. Have students scatter a handful of toothpicks, paper clips, counters, or other small objects on a sheet of grid paper. Have them estimate the number any way they can. Talk about strategies, such as visualizing rows or columns, breaking the area into smaller sections, or visualizing patterns. You can do this on the overhead projector as a class activity.

◆ Students may enjoy "going dotty" by making dot pictures for classmates to estimate. They can use a pencil eraser dipped into thinned paint to print dots on plain paper. They can fill envelopes with the paper dots that collect inside a 3-hole punch. They can even scatter grains of rice on paper to attempt a similar investigation.

◆ Discuss the analogy between this kind of estimation and crowd estimates made from helicopters or aerial photographs.

◆ Display reproductions of works by Georges Seurat and Roy Lichtenstein, artists who used dots to create pictures.

 *Estimates will vary.*

Name _____

How Many in a Jar?

Examine the container your teacher has filled, but do NOT open it. Estimate how many objects it holds. Write your estimate.

Now try again. Here are some tips:

◆ Imagine dividing the container into *handfuls.* How many objects can you imagine in each handful? Should you change your estimate?

◆ Estimate how many objects in one *layer.* Think about how many layers there are in all. Should you change your estimate?

◆ How else could you think about the objects in the container to help you make a better estimate?

Now open the container and count. How close was your best estimate?

Name _____

Dot, Dot, Dot...

Look at the dots in the frame. Estimate how many there are. What strategy helps you make the best estimate?

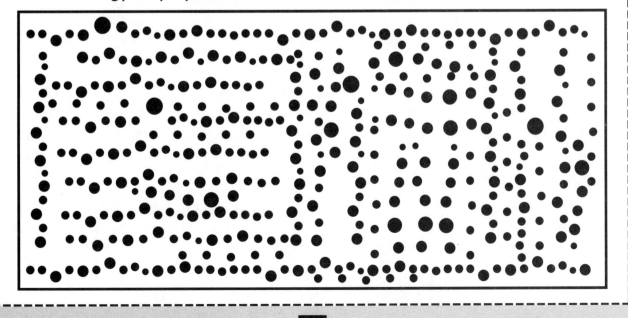

Shoefuls

Connections: number sense; measurement; spatial reasoning

Teaching Tips:

◆ Talk about ways to make a reasonable estimate. Fill a baby shoe with centimeter cubes (or another kind of object) as students count along with you. Ask: *How can you use what you know about the capacity of this shoe to help you estimate the capacity of your shoe?*

◆ Suggest that students work in groups. They might use one person's shoe as a standard upon which to base other estimates. What factors do they consider in filling the shoe? Do estimates improve over the course of the investigation? Groups should talk together to establish a common set of guidelines to use as they work. Provide time for them to share results. What generalizations can be made about shoefuls?

◆ This investigation can be done many times, using different styles of shoes or fillers of different sizes and shapes. Help students graph their results.

✔ *Estimates will vary.*

Bulb Life

Connections: time; probability; functions; number sense

Teaching Tips:

◆ Bring in several light bulb packages that provide information about average bulb life. Provide this information to students as a lead-in to the investigation. Ask: *If you know that this bulb is expected to last 1,000 hours, when will you have to replace it?* It may help students to have access to calendars as they make their estimates.

◆ Help students estimate the amount of time lights are typically on in particular rooms at home, in a classroom, in a street lamp, and so on. Discuss factors that affect bulb life, such as the number of times it is turned on and off (each successive on/off shortens its life), the kind of bulb (incandescent, halogen, fluorescent, energy saver, etc.), whether it is wired to a dimmer, etc. It might help to talk to the school custodian about the kinds of bulbs used in classrooms.

◆ Follow up by having students make predictions about when they will have to change a bulb at home or in the classroom. Keep the estimates. When it is time to replace a bulb, check the estimates. Do any match to what really happened?

✔ *Estimates will vary.*

Name_____

Shoefuls

How much can your shoe hold? It's an odd shape and you can't easily see inside it, but it's fine to use for estimating.

Take off one shoe. Feel around inside it. Estimate how many of each kind of object will fit inside it without overflowing.

- ◆ **centimeter cubes**
- ◆ **crayons**
- ◆ **paper clips**
- ◆ **jelly beans**
- ◆ **place value block tens (longs)**

Then try it! How do your estimates compare with the exact numbers? With those of a classmate? How can you account for variations?

Name_____

Bulb Life

A bulb rated to last 1,000 hours should last about 1,000 hours. But how long is that? When would you expect to replace it if you installed such a bulb today in one of the following:

- ◆ **your classroom?**
- ◆ **your bedroom?**
- ◆ **your family room or living room?**
- ◆ **your basement or garage?**
- ◆ **a street lamp?**

Work with your group to make these estimates. It may help to use a calendar.

☐ To Estimate or Not?

Connections: number sense; visual/spatial reasoning; measurement

Teaching Tips:

◆ Before students work through the statements on the page, ask them to think of times in their own lives when they estimate—i.e., how long it takes to make a phone call to a friend, how long to do the dishes, how much money to bring to the movies, etc. Discuss the difference between an exact answer and an estimate, and when one may be more suitable than the other.

◆ Feel free to adjust any of the statements given to make them more applicable to your students' experiences.

◆ Expect students to have different answers. Encourage discussion and comparison of strategies and reasoning. You might challenge students to write in their math journals to explain how and why they made any particular decision.

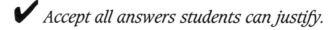 *Accept all answers students can justify.*

☐ Let's Be Reasonable

Connections: number sense; measurement; logical reasoning

Teaching Tips:

◆ If you wish, you may create your own fill-in-the-number statements, modeled after the ones given, that may be more suitable to your students' experiences and interests.

◆ Have students make up 5 more incomplete statements like these. They can exchange them with a classmate to complete.

◆ These statements work well to motivate discussion. You might want to generate a similar set as an introductory activity for a family math evening.

◆ For statement 6, encourage students to suggest a variety of responses, such as 7 more hours sleeping than riding the school bus.

 Estimates will vary.

Name_____

To Estimate or Not?

Read each statement. Decide whether you need an exact answer or if an estimate makes more sense. Explain your decision.

1. Your parents need to know the size of your bedroom window so they can order blinds for it.

2. You want to know how much pepper to put in the stew.

3. It's your first time at bat in a softball game. You want to know how far from the plate to stand.

4. You must decide how much water to give your house plants.

5. You are mailing a package to Brazil. You need to know its weight to know how much postage to put on it.

Write 5 more statements like these. Ask a classmate to decide.

Name_____

Let's Be Reasonable

Fill in sensible numbers. Tell how you made your choice.

1. Our school has about _____ students.

2. A sundae with everything on it costs about _____.

3. Washington, D.C. is about _____ miles from Tucson, Arizona.

4. A van is about _____ feet long.

5. A duck weighs about _____ pounds.

6. An elevator holds about _____ passengers.

7. You usually spend about _____ more hours a day _____ ing than you do _____ ing.

⬜ Numbers in the News

Connections: number sense; percent; measurement

Teaching Tips:

◆ Collect a variety of headlines from your local newspaper that contain numerical data. Present them to students and talk about how the numbers were obtained. Talk about the benefits of estimates in some situations *(easy to understand, easy to read and remember)* as compared to the benefits of exact numbers *(more precise)*.

◆ Allow time for discussion and disagreement. Help students respect differing opinions, and encourage them to support their ideas as clearly as possible.

◆ Students may be puzzled by statements that could be exact numbers or estimates, such as statements 3, 6, and 7. Help them justify either choice.

◆ For a homework assignment, ask students to find other examples of numbers in the news. You might make a bulletin board display of the headlines, challenging students to classify them as estimated or exact numbers.

✔ *Accept any responses students can justify.*

⬜ Over or Under?

Connections: number sense; time; logical reasoning

Teaching Tips:

◆ Before students begin, be sure they understand the meaning of overestimating *(guessing high)* and underestimating *(guessing low)*. You might offer real-life examples of times when it is better to overestimate *(i.e., deciding how much time to allow to get to the theater and wait on lines for tickets and popcorn without missing the beginning of the film)* and underestimate *(i.e., when you want to convince your parents that you only have a few more minutes of homework to do)*.

◆ Invite a school worker to your class to discuss the kinds of estimates they use in their jobs. For instance, the cafeteria cook estimates the number of lunches to prepare daily, and probably overestimates so there will be enough food for everyone. The librarian may estimate the number of books that circulate each week, and so on. As a follow-up, ask students to talk with family members about home estimation.

✔ *For #2, an underestimate is better because you cannot remove excess salt once added.*

Numbers in the News

Which headlines are estimates? Which give exact figures? Explain.

1. **250,000 AT PARK CONCERT**
2. **MAN EATS 25 HOT DOGS!**
3. **LOTTERY PRIZE REACHES $11 MILLION**
4. **STATE POPULATION UP 3%**
5. **PLAYOFF GAME DRAWS 52,530 FANS**
6. **40-ACRE ANCIENT WALLED CITY FOUND**
7. **10 MILES OF SUBWAY TRACK UNDER CONSTRUCTION**
8. **SURVEY FINDS 1 IN 4 STUDENTS HATE PEANUT BUTTER**

Make up 5 more headlines. Ask a friend to pick the estimates.

Over or Under?

Read each statement. Decide whether an *over*estimate or an *under*estimate makes more sense. Explain your decision.

1. You're estimating expenses to determine how much money to bring with you to the street festival.

2. You're estimating how much salt to add to the soup.

3. You've got a list of chores to do by lunch time. You estimate how long each will take, to decide how many you'll do and how many to assign to your cousin Norman.

4. Shalini will pick you up from a rehearsal. You estimate how long the rehearsal will last, so you can tell her when to get there.

□ Fit the Estimates

Connections: number sense; measurement

Teaching Tips:

◆ The statements in this activity are open-ended to encourage students' creativity and divergent thinking. You may find that some students will need clues to help them focus their thinking. For instance, for statement 1, you might say, "Think about when 200 years ago was. What are some things that happened then?" For statement 3, you may need to help students find a familiar benchmark, such as the number of stories in a well-known building in your community.

◆ You might have students work in groups to brainstorm as many different responses to each statement as they can in the time allowed.

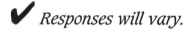 *Responses will vary.*

□ Around a Pound

Connections: visual/spatial reasoning; measurement

Teaching Tips:

◆ Gather an assortment of 1-pound weights before beginning this investigation. They can be official weights you might borrow from the science teacher, or common items, such as a 1-pound can of fruit, bag of rice, box of pasta, block of cheese, or anything you can conveniently collect. Objects may have different sizes and shapes, as long as they weigh 1 pound.

◆ You need a balance scale or a platform scale that can weigh to the nearest pound. Gather a variety of small objects groups of students can examine, or allow them to look around the classroom or in their desks for things they can group and compare.

◆ Provide a weight to each group of students. Encourage them to act as equal-arm balances and try to identify objects of equivalent weight.

◆ As an at-home activity, ask students to accompany a parent to the grocery store. Have them try to estimate the number of food items to the pound, such as apples, potatoes, onions, bananas, and so on, or estimate the weight of the particular amount of these foods their parents purchase.

 *Estimates will vary.*

Name _____

Fit the Estimates

For each statement, think of something that fits.

1. It's about 200 years old.
2. It weighs about an ounce.
3. It's about 10 stories high.
4. It's about a mile from school.
5. It takes about a half-hour to do.
6. It costs about $1,000.

7. It happens maybe twice a year.
8. It can hold about 25 potatoes.

For statements 1–7, think of something that's a little *more.*
Then think of something that's a little *less.*

Name _____

Around a Pound

Examine the 1-pound object your teacher has prepared. Use what you learn from handling it to help you do the investigations below. Check your results on a scale.

1. Name 3 things that each weigh about a pound.

2. How many crayons are in a pound? How many index cards? How many calculators?

3. Put together a collection of things with a total weight of about a pound. Then put together another collection, using *more* things.

4. Make a new collection with at least 3 objects whose total weight is about a pound. Have your partner do the same. Who's closer?

? ☐ Just a Minute

Connections: time

Teaching Tips:

◆ You might begin this investigation by asking everyone in the class to cover their eyes. Say "Go" and time 1 minute. Students should silently raise their hands when they think 1 minute has passed. Keep track of guesses that are under a minute, over a minute, and exactly 1 minute. Talk about how students estimated, then try again.

◆ As students brainstorm events that take about 1 minute, help them broaden their lists by guiding them to consider kitchen activities, school activities, sports activities, chores, entertainments, and so on.

◆ Discuss the effect one's feelings or attitudes have on estimates of time. For instance, talk about times when 1 minute seems to pass in a flash, such as during an exciting game with a friend; other times a minute can seem like forever, such as when waiting for the bus on a blustery morning.

◆ Talk about expressions people use, such as "I'll be done in 5 minutes," "Give me a minute," and so on. Have students list some of these expressions and then interview adults to find out what they *really* mean when they say them.

 *Estimates will vary.*

☐ Reading Rate

Connections: time; proportional reasoning; number sense; visual reasoning

Teaching Tips:

◆ For this investigation, students need a stopwatch or a clock with a second hand.

◆ To determine reading rate, students can use the given paragraph, or they can select paragraphs from books they are reading. You may want to provide paragraphs of comparable length or set up parameters for the paragraphs they choose, such as a minimum number of lines or words.

◆ Have students record their estimates and actual times in a table. Have them try a variety of rate estimates, such as the time it takes to read a paragraph, a page, or a chapter. Extend the investigation to explore how long it takes to follow a set of directions, find their family's listing in the local telephone directory, fill out an application, read a favorite section of the newspaper, or do a word puzzle.

 Estimates will vary.

Name _____

Just a Minute

How accurate is your internal clock? Do you know how long a minute lasts?

1. Work with a partner guessing how long a minute lasts. You'll need a stopwatch or a clock with a second hand. One of you will act as the timer, the other as the estimator.

2. The timer says "GO!" and the estimator, without looking at the time, signals when he or she thinks a minute has passed. The timer tells whether the estimate was over or under a minute.

3. Try it 3 times. Do your estimates improve?

4. Take turns guessing.

Talk with your partner. List all the everyday activities you can think of that take about a minute to do. How about activities that take about 5 minutes? About 10 minutes? About a half-hour? Be prepared to share your list with the class.

Name _____

Reading Rate

How long will it take to read this paragraph? Guess, then find out.

Estimate how long it will take you to read a paragraph or a whole page in a book you are reading. Then try it to find out.

◆ At that rate, about how long would it take to finish the book?

◆ What other factors might help you fine-tune this estimate?

◆ Do you read fiction and nonfiction at the same rate? What about newspapers or magazines vs. books? Explain.

Meg invited me to her house. She was eager for me to meet Tracy, whom she was sure I'd like. "Why not?" I thought. So we made our way to her house and went to the basement. Something seemed a little odd, because it didn't look much like a family room. It was dark, quiet and had a musty smell. Meg was trying to hide a smile, she watched me closely. No wonder! Tracy was a bigger surprise than I ever imagined, and a longer one, too. Tracy was a snake, a 15-foot-long boa constrictor. "She's very sweet," Meg crooned as she picked her up and held her out to me.

With EEEEs

Connections: *number sense; proportional reasoning; statistics*

Teaching Tips:

◆ Open this investigation by writing a sentence on the chalkboard and having students estimate how many letters it contains. Then have them estimate how many of the letters are vowels. Have volunteers confirm the estimates.

◆ Ask groups to select a page they can examine to estimate the number of **E**s it has. You may want all students to examine the same page, or you may prefer to have them work with different pages to generate a variety of data. Help students establish a benchmark to help them make more accurate estimates. For example, they can count the number of **E**s in one line, sentence, or paragraph, then multiply that number by the approximate number of comparable-sized sections in the entire passage.

◆ If you have access to books in other languages, you might try a similar investigation in another language to see what relationships emerge, if any.

✔ *Estimates will vary.*

Word Finder

Connections: *number sense; visual reasoning; proportional reasoning*

Teaching Tips:

◆ One strategy students might use to estimate page numbers is to estimate what fraction of the alphabet contains the initial letter, then find that fraction of the number of pages in the book. For example, a word that begins with **M** is about half way through the alphabet, so it might appear about half way through the pages in the dictionary. Whatever strategy students use, encourage them to explain it to the class.

◆ Try a related investigation in which you give a page number, say 297, and challenge students to name a word they might find on that page. To do this, they might use a strategy that reverses the one described above.

◆ You can repeat this investigation, using different word lists or different dictionaries.

✔ *Estimates will vary.*

With EEEEs

People who play *Scrabble*™ or "Hangman" or watch *Wheel of Fortune* on TV know that **E** is the most commonly used letter in the English language. How many **E**s do you think you will find on one page of the book you are reading? Guess, then investigate to find out. You can give your estimate as a range.

HINT: You may want to establish a *benchmark* to help you make a better estimate by finding the number of **E**s on <u>this</u> card.

Compare the estimates and actual results with your classmates. What do you notice? What might account for variations?

Name _____

Word Finder

Looking up a word in the dictionary involves more than knowing alphabetical order. It takes estimation, too.

◆ Use a dictionary. How many pages does it have in all?_____

◆ If you opened to the middle page of your dictionary, what letter do you think the words there would begin with?_____

Estimate the page number where you might find each of the following words in your dictionary. Then look them up to confirm the estimates.

	Estimate	Actual
novella		
queue		
maxim		
apogee		
rhinal		
gaggle		

	Estimate	Actual
cudgel		
indolent		
smarmy		
yegg		
lunette		
joist		

☐ Checkout Counter *(student page on 35)*

Connections: number sense; money

Teaching Tips:

◆ Before students try this investigation, you may want to review some computational estimation strategies (see Appendix, pages 73–79).

◆ Bring in some grocery store receipts. Remove the totals, but code the receipts in some way so you have an answer key. Display them on a bulletin board or in a folder in the Estimation Corner. Invite students to try some more grocery estimation on their own.

✔ *Sensible estimates: 1. $15; 2. $17; 3. $20; 4. $23*

☐ Fred's Farm Stand *(student page on 36)*

Connections: number sense; money; measurement

Teaching Tips:

◆ Have students work in pairs so they can talk about their estimation methods. You may want to challenge groups to find as many different answers as possible for problem 1.

◆ If necessary, point out the different units used for items on sale at Fred's Farm Stand *(pound, pint, bunch, each)*. Remind students to label answers with the appropriate unit.

✔ *1. Answers may vary; 2. about 4 lb; 3. about $4; 4. about $6.40; 5. strawberries; 6. no; 7. about $2.50*

☐ Pet Store Sale *(student page on 37)*

Connections: number sense; percent

Teaching Tips:

◆ Review benchmark percents, especially 25% and 50%.

◆ Have students use local store ads or sale flyers to estimate other sale prices based on percent-off sales.

✔ *1. $14, $5; 2. $75, $26; 3. $5.50; 4. all but* Power Puppies, Cats on Mats; *5. yes; 6.* Cats on Mats, *$5*

Checkout Counter

Estimate the total of each of the following supermarket receipts. Use any techniques you know. Compare answers and strategies with your group.

1.

produce	$.89
produce	$1.77
grocery	$2.25
dairy	$2.99
dairy	$3.19
grocery	$2.45
grocery	$.79
bakery	$1.15

2.

meat	$3.79
meat	$2.98
dairy	$1.59
dairy	$3.29
produce	$.79
dairy	$.65
produce	$1.19
dairy	$.89
produce	$1.39

3.

dairy	$2.95
grocery	$4.29
grocery	$.98
grocery	$1.59
produce	$1.29
grocery	$2.85
produce	$.69
dairy	$1.09
bakery	$2.25
bakery	$1.69

4.

bakery	$.45
dairy	$.80
dairy	$.99
produce	$1.47
produce	$2.01
grocery	$5.39
grocery	$4.60
grocery	$1.98
grocery	$2.59
produce	$1.29

◆ What estimation strategy would you use to get the quickest estimate? Explain.

◆ Which strategy would you use to get the most precise estimate? Why?

Name _____

Fred's Farm Stand

It may be corny, but at Fred's farm stand, he only sells fruits and vegetables that are red. You won't see lemons there, nor oranges, and certainly never lettuce or broccoli.

Use estimation and the price list to answer each question. Talk about the strategy you used.

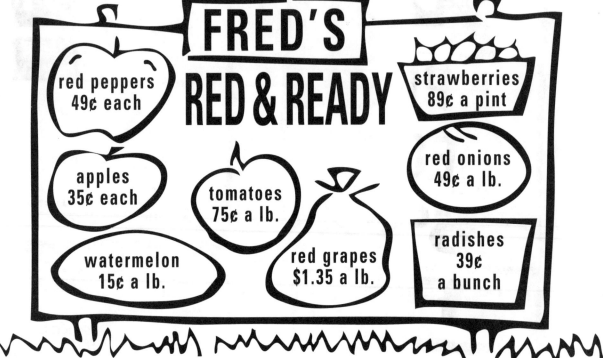

FRED'S RED & READY

red peppers 49¢ each

apples 35¢ each

tomatoes 75¢ a lb.

watermelon 15¢ a lb.

red grapes $1.35 a lb.

strawberries 89¢ a pint

red onions 49¢ a lb.

radishes 39¢ a bunch

1. You have a dollar. What might you buy?

2. How many pounds of red grapes can you buy with $5? Explain.

3. About how much would an apple, 2 pounds of tomatoes, and a 14-lb watermelon cost?

4. Approximately what would 3 pints of strawberries, 2 pounds of red onions, and a 2-lb bunch of red grapes cost?

5. What would you guess will cost more, 10 bunches of radishes or 5 pints of strawberries?

6. Can you buy 4 pounds of red onions, 4 red peppers and 4 apples for $5.00? Explain.

7. Suppose you buy 3 bunches of radishes, 3 pounds of tomatoes, and 3 pounds of red grapes, and you pay with a $10 bill. About how much change will you get?

8. Make up a vegetarian estimation problem for a classmate to solve. Swap questions.

Name_____

Pet Store Sale

Things are jumping at Leaping Lizards Pet Shop. All pet exercise videos are discounted. Use estimation to answer the questions below.

1. About what would it cost to buy a copy of *Gerbil Gymnastics*? About how much would you save?

2. About what would 3 copies of *Cats on Mats* cost? About what would the savings be?

3. About how much more would a copy of *Dog Dancercises* cost than a copy of *Turtles of Steel*?

4. Which videos are under $25 with the discount?

5. Can you buy a copy of *Power Puppies* and 2 copies of *All-Pet Workout* with $60? Explain.

6. Suppose Leaping Lizards holds a special 50%-off sale on all *Cats on Mats* videos. Which is cheaper, a *Dog Dancercises* video or *Cats on Mats*? About how much cheaper?

Home Wise

Connections: number sense; visual/spatial reasoning; measurement

Teaching Tips:

◆ To help students make reasonable estimates for the various objects in their homes, have them imagine a trip through the house. They can visualize each room to picture every feature they want to estimate.

◆ Students can use the same kind of visualization techniques to help them estimate the number of steps between parts of their home. You may prefer to have students estimate in terms of standard measures, such as feet or meters.

◆ Many students will need to try some length estimates in class first. They can estimate, then count the number of steps (or feet or meters) from their desk to the class pencil sharpener, from the classroom door to the teacher's desk, and so on. Establishing visual benchmarks is a useful estimation technique, which can be applied to any investigation involving length or distance.

◆ You may want students to graph or tally the results when they make exact counts at home. Encourage them to share the techniques they used to make their estimates, to analyze how close their estimates were to the exact counts, and to explain why some estimates may have been closer than others. Invite groups to compare their findings.

✔ *Estimates will vary.*

Name_____

Home Wise

How well do you know your own home? This investigation may help you to find out.

Think about your home, room by room. Estimate the total number of each of the objects listed below. On a seperate sheet, add other items to estimate. Tonight for homework, find the exact count. Tell whether you *over*estimated or *under*estimated.

ITEM	ESTIMATE	ACTUAL	OVER or UNDER?
electrical outlets			
clocks			
windows			
doors			
drawers			
items on the wall			
coat hangers			

Which was the hardest estimate for you to make? _____
Why? _____

Now think about another way to describe your home. Make a table like the one above. This time, estimate the number of steps—

- ◆ from the foot of your bed to the nearest light switch;
- ◆ from your bedroom to the bathroom;
- ◆ from the kitchen to wherever you dispose of garbage;
- ◆ from the front door to the door to your room;
- ◆ from your front door to the street;
- ◆ around the outside of your home.

At home, verify your estimates. Which was the closest?

❓ Wallpapering

Connections: geometry; measurement; visual reasoning; proportional reasoning

Teaching Tips:

◆ Ask students to estimate the dimensions of a dollar bill. Then have them use a ruler to find its actual dimensions.

◆ Students will suggest many ways to approach this investigation. Some may want to measure the surface area of the walls in the classroom. Others may be more comfortable visualizing a small section of the wall papered with dollar bills and extrapolating from there. Encourage a variety of approaches.

◆ As a follow-up activity, you might ask students to try the same investigation at home for their own bedroom.

✔ *Estimates will vary; a dollar bill is about 6 in x 2 1/2 in.*

Pets on a Plane

Connections: geometry, measurement; visual/spatial reasoning

Teaching Tips:

◆ Ask students to describe pet carriers they have seen. Have them explain how the size of the carrier compares to the size of the pet it is designed to hold. *(large enough for some movement, but not larger than necessary)*

◆ Students may need to research the size of less familiar animals, such as the llama or the box turtle. An encyclopedia or other reference book on animals may help.

◆ Students who especially enjoy this investigation may want to follow it up by designing carriers for exotic pets.

✔ *Estimates will vary.*

Name _____

Wallpapering

Imagine that a very rich, very fussy, and very generous person wants to improve the appearance of your classroom walls. Also imagine that this person has offered to replace the paint job with wallpaper—made of dollar bills!

This person will supply the bills, but you must do the work. How many dollar bills would you estimate it takes to wallpaper all the walls of your classroom? On a seperate sheet describe a strategy you can use to find out.

Name _____

Pets on a Plane

When Byron travels, he likes to take one or more of his pets along with him. Think about how a pet carrier looks. Think about approximate sizes and shapes of the pets listed below. Then estimate the dimensions for a carrier for each pet passenger. On a separate sheet draw a sketch to help you; label the dimensions.

- ◆ **Boris the box turtle**
- ◆ **Carmine the cat**
- ◆ **Alana the llama**
- ◆ **Horace the horse**
- ◆ **Pat and Pete, the Pekinese puppies (they travel together)**

Each of the three investigations that follow involves the use of maps. Note that the teaching tips at the bottom of this page refer to all of them.

☐ Cruising Carr County (student page on 43)

Connections: map reading; visual reasoning; number sense

✔ *1. 10 mi; 2. 35 mi; 3. 60 mi; 4. 50 mi; 5. 15 mi; 6. Answers will vary; 7. 45 mi.*

☐ Getting There (student page on 44)

Connections: map reading; visual reasoning; number sense; time; proportional reasoning

✔ *1. 2 h; 2. 2 h; 3. Estimates will vary; 4. car; 5. moped; 6. Answers will vary.*

☐ At the Science Center (student page on 45)

Connections: map reading; visual reasoning; number sense; time; proportional reasoning

✔ *Plans will vary.*

Teaching Tips:

◆ Ask students to work in pairs so they can talk about the maps together and discuss possible approaches they can use to solve the problems.

◆ Some students may use string to help them estimate distances, especially on curves.

◆ Be sure students understand the meaning of the expression "as the crow flies" (shortest distance between 2 points), which they will encounter in *Cruising Carr County*.

◆ Review how to use a map scale.

◆ For *At the Science Center*, allow time for students to work out their plans, share them, and revise them as they see fit. You might choose to allow students' interests to determine the amount of time they spend in an exhibit.

◆ If there is a large museum near you, you might substitute actual floor plans or brochures so that students can try a similar investigation based on real-life data.

Name _____

Cruising Carr County

On the map below, Pistonville is 15 miles from Wagon Station.
Use this information to answer the questions.

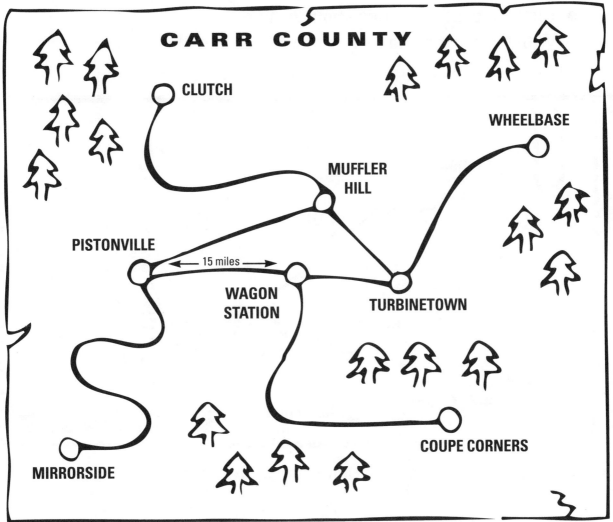

1. About how far is the drive from Wagon Station to Turbinetown?

2. About how far is the drive from Pistonville to Mirrorside?

3. About how far is the drive from Wheelbase to Coupe Corners?

4. Estimate the distance of the shortest trip from Muffler Hill to Coupe Corners.

5. About how much farther is the shortest route from Turbinetown to Mirrorside than the shortest route from Wagon Station to Clutch?

6. Describe a drive that is about 35 miles. Describe one that's about 60 miles.

7. As the crow flies, about how far is it from Clutch to Coupe Corners?

Name _____

Getting There

Kim's family has three vehicles—a car, a moped, and a tractor.

◆ Kim drives her car at a speed of about 50 miles per hour (mph).

◆ On the moped, she goes about 25 mph.

◆ With the tractor, she can go only about 15 mph.

Use Kim's driving speeds and the map to answer the questions.

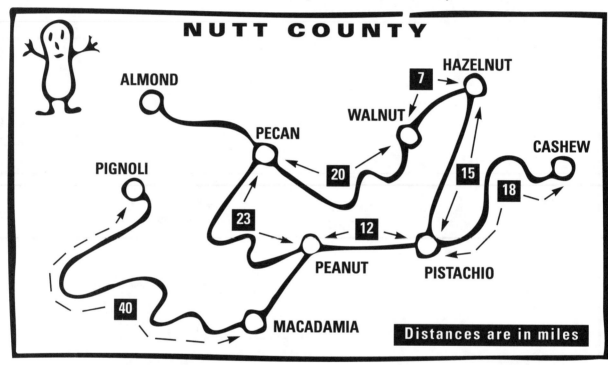

NUTT COUNTY

ALMOND PECAN PIGNOLI PEANUT MACADAMIA HAZELNUT WALNUT CASHEW PISTACHIO

7   20   23   12   15   18   40

Distances are in miles

1. About how long would it take to drive the moped from Peanut to Pignoli?

2. About how long would it take to drive the tractor from Almond to Walnut?

3. Describe a drive that would take about a half-hour by car. Describe one that might take about an hour.

4. Kim drove from Hazelnut to Macadamia by way of Walnut. The trip took just over an hour. Which vehicle was she driving?

5. Kim drove from Pecan to Cashew via Peanut. The trip took a little over 2 hours. Which vehicle was she driving?

6. Kim drove the tractor for about 5 hours one day. Describe a trip route she might have taken.

Name _____

At the Science Center

Spend some time in a hands-on science center.

Look at the map below. Note the suggested times to allow to explore each exhibit. Some popular ones have waiting times.

Use all the information to plan a 3-hour stay. Write a schedule that shows the exhibits you'll visit, the order in which you'll visit them, and the amount of time you'll spend in each. Remember to allow time to walk from one exhibit to another. After 3 hours, you must meet your teacher back at the entrance.

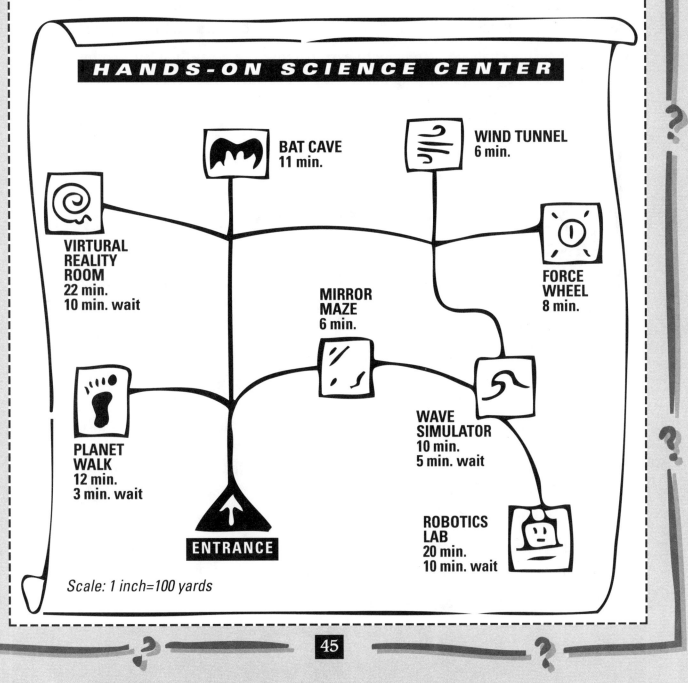

HANDS-ON SCIENCE CENTER

BAT CAVE
11 min.

WIND TUNNEL
6 min.

VIRTURAL
REALITY
ROOM
22 min.
10 min. wait

FORCE
WHEEL
8 min.

MIRROR
MAZE
6 min.

PLANET
WALK
12 min.
3 min. wait

WAVE
SIMULATOR
10 min.
5 min. wait

ROBOTICS
LAB
20 min.
10 min. wait

ENTRANCE

Scale: 1 inch=100 yards

Get the Lead Out

Connections: strategy formulation; visual reasoning; measurement; proportional reasoning

Teaching Tips:

◆ This investigation requires students to plan a strategy that resembles designing an experiment. It may help for students to work in groups of 3 or 4 so they can pool their ideas and plans and get a wider range of data to check their estimate.

◆ Discuss the parameters of the task. Some questions to consider: *If a pencil point breaks off right away, can the student sharpen the pencil and start again? Do we want to agree on a particular kind of paper? On the amount of pressure to apply?*

◆ When groups finish the investigation, have them compare results as a whole class. You may want to make a class graph of the outcomes to conclude the investigation.

◆ As a follow-up, challenge students to extrapolate from what they learned in their investigation to estimate how long a line they could draw with an entire pencil.

 *Estimates will vary.*

How Likely Is It?

Connections: probability; fractions; logical reasoning

Teaching Tips:

◆ Review the language of probability, such as *probably, certain, impossible, chance, unlikely,* and so on. Brainstorm with the class to list all the words students know that relate to probability.

◆ As necessary, review the fractional benchmarks used to describe probability: 0 means no chance of an event occurring; 1/2 means equal chance that an event will or will not occur (i.e., "50-50 chance"); and 1 means an event is certain to occur.

◆ Some students may have difficulty with this kind of estimation. For instance, although an event may not really have a 0 probability, 0 may be the most reasonable estimate when the choices are 0, 1/2 or 1. Allow sufficient time for students to explain their reasoning, discuss variations in responses, and offer counterexamples.

◆ Adjust the days given in the statements to fit your schedule, if necessary.

◆ Have students make up other probability statements of their own. Have them exchange papers with classmates so they can evaluate the probability of each statement.

✔ *Estimates will vary.*

Name _____

Get the Lead Out

Can your pencil keep up with all the writing you do?

Sharpen your pencil to a fine point. How long a line do you think you could draw with it until the graphite is so worn away that you cannot draw any more?

Work with a partner. Design a strategy you could use to make a reasonable estimate. Try your strategy. Describe what you did. Compare your results with those of your classmates.

What factors can affect the length of the line you can draw?

Name _____

How Likely Is It?

Think about the likelihood of each event below. Estimate the probability of each by writing 0, 1/2, or 1. Be prepared to explain your answer.

1. More than 3 students in your class will be absent on Tuesday.

2. You will be interrupted by a fire drill later this week.

3. Everybody will like the food served in the cafeteria tomorrow.

4. Your teacher will give you a note to take home today.

5. You will use your pencil tomorrow.

6. At least half your class will wear sneakers on Friday.

7. You will have a substitute teacher next week.

8. Someone in your class will hiccup tomorrow during reading.

☐ **Shading Shapes** *(student page on 49)*

Connections: *percent; visual reasoning; measurement*

Teaching Tips:

◆ Review benchmark percents, such as 25%, 50%, 75%, 33 1/3%, 66 2/3%, 20%, and so on. You may want to review the connection between these key percents and fractions or decimals.

◆ You might want to begin this investigation by having students shade 50% or 25% of common polygons drawn on grid paper so they have a way to verify their estimates.

◆ Suggest to students that they use pencil so they can erase or adjust their shadings as necessary. They can use grid paper or tracing paper to verify their estimates.

✔ *Estimates will vary; check students' work.*

☐ **In the Shade** *(student page on 50)*

Connections: *percent; visual/spatial reasoning; measurement*

Teaching Tips:

◆ You might view this investigation as an extension of the preceding one. Here, students must determine a reasonable percent to use to describe the shaded part of the given figures. Begin by showing some concrete objects that have a part shaded or designated in some way. For instance, you might show a jar partly filled with sand or a sheet of paper partially covered. At first, have students compare the shaded or filled region to 50%, 25%, 75%, or 100%. Then challenge them to make more specific estimates.

◆ One strategy students can use to check their estimates for the shapes on the page is to trace each figure onto grid paper and count squares and partial squares.

◆ Students with weakness in spatial reasoning may be more successful if you assign this page after they have had more time exploring in a more hands-on way.

✔ *Estimates will vary; check students' work.*

Name _____

Shading Shapes

Estimate to shade each figure as described. Be prepared to describe your strategy. Think of a way to check your estimate.

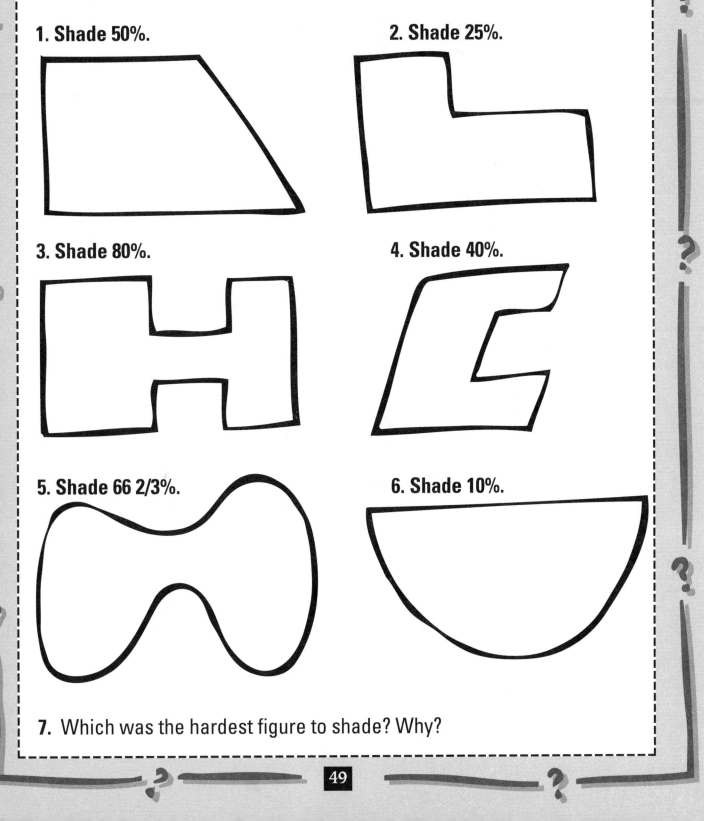

1. Shade 50%.

2. Shade 25%.

3. Shade 80%.

4. Shade 40%.

5. Shade 66 2/3%.

6. Shade 10%.

7. Which was the hardest figure to shade? Why?

Name _____

In the Shade

Examine each drawing. Estimate the percent that is shaded. Explain how you made your estimate.

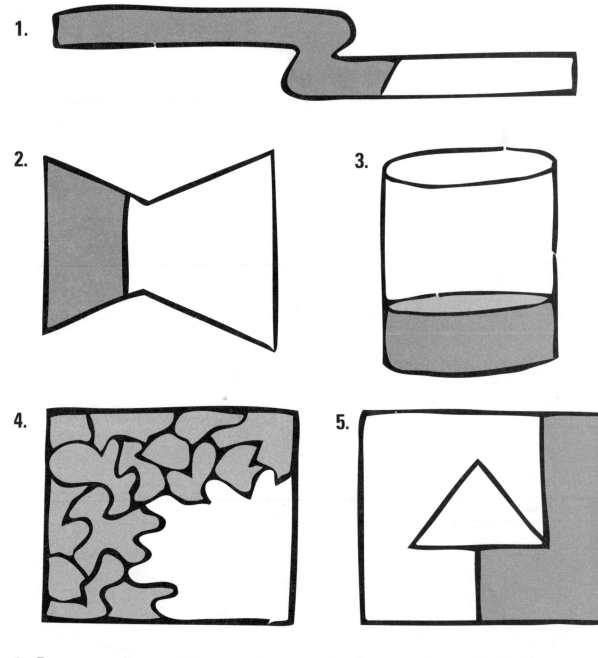

1.

2.

3.

4.

5.

6. Draw a trail map. Place a hiker somewhere on the trail. Challenge a classmate to estimate the percent of the trail the hiker has walked.

PART 3
Take Longer

How Far Is the Flick? *(student page on 53)*

Connections: measurement; spatial reasoning; data collection

Teaching Tips:

◆ Agree on a level of precision for estimates based on the ability and experience of the estimators—to the nearest foot, 1/2-foot, inch, etc. You can vary the investigation by using different objects or ways to move them (i.e., roll, push, "tiddlywink", blow).

◆ As a follow-up, challenge students to estimate the distances between classroom objects or locations—i.e., how far is it from their desk to the pencil sharpener, to the sink, etc.

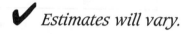 *Estimates will vary.*

Cupfuls *(student page on 54)*

Connections: measurement; spatial reasoning; data collection

Teaching Tips:

◆ Gather different sizes (and shapes) of paper or plastic cups, enough so that each group can have one of each. Label same-size cups (by letter, color, or number) for ease of comparing estimates. Provide large containers (pitchers, pots, jugs, large jars, etc.).

◆ If it is difficult to provide access to water for this investigation in your class, alter the task by having students use sand; or assign the investigation to be done at home.

 *Estimates will vary.*

Area Mark-Off *(student page on 55)*

Connections: measurement; spatial reasoning; geometry

Teaching Tips:

◆ Review area of rectangles (l x w). Have students use geoboards, dot paper, or grid paper to find as many different rectangles as they can that have an area of 12 square units.

◆ Students can use a floor tile, piece of paper, or other rectangular figure as a benchmark. For step 4, have students work in a different part of the room, or remove visual cues.

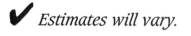 *Estimates will vary.*

Name_____

How Far Is the Flick?

Work with a partner on a distance estimation investigation.

You'll need—

◆ a small object (coin, button, centimeter cube, paper clip, etc.)

◆ a measuring tool (ruler, tape measure, meter stick, etc.)

◆ an open area in your classroom or in the hall

◆ tables like this to record your results—

Flick	Estimate	Actual Measure	Comments
1			
2			
3			
4			
5			

Here's what to do:

1. Flick the object from an imaginary starting line on the floor. How far did it go? Record your estimate.

2. Measure the actual distance. Record it. How close was your estimate?

3. Flick again. Estimate and record. Measure and record. Compare. Then repeat the process three more times.

4. Now choose a different object. Flick it. Estimate and record the distance. Measure and record the actual distance. Compare. Try this five times.

5. Discuss the results of your investigations with classmates. How close were your estimates? Did they improve with practice? Explain.

Name_____

Cupfuls

Try a capacity investigation, a cupful at a time!

You'll need—

- ◆ several paper or plastic cups of different sizes
- ◆ a large container
- ◆ a sink
- ◆ paper towels or a sponge, just in case!
- ◆ a table like this one—

Cup	Estimated Number to Fill Large Container	Actual Number	Comments

Here's what to do:

1. Examine the different cups and the large container. Think about how they compare in terms of the amount of water they can hold.

2. Choose one of the cups. Estimate how many of cupfuls that size it would take to fill the large container with water. Record your estimate.

3. Now use that cup to find out! Count as you fill 'er up. Record the total.

4. Repeat the investigation using each of the other cups. Use what you learned about the first cup to help you make your estimates for the other cups.

Think about your estimates. Describe the strategies you used. How does the accuracy of your later estimates compare with your first one? Explain.
Talk about it with your classmates.

Name _____

Area Mark-Off

What does a region with an area of 12 square feet look like?

Could you park a car in a region that size? Would your teacher's desk fit in it? Is it large enough for you to stand in and practice juggling? Try an area investigation to find out.

You'll need—
- ◆ tape or string to outline the region
- ◆ a measuring tool (ruler, tape measure, meter stick, etc.)
- ◆ open area (in the classroom or outside)
- ◆ paper

Here's what to do:

1. Use tape or string to mark off a rectangular region on the floor that you estimate to be about 12 square feet.

2. Measure to check your estimate. How long is your rectangle? How wide is it? How close did you come to marking off 12 square feet? Did you *over*estimate or *under*estimate? On a piece of paper, draw a small-scaled picture of your rectangle that gives its dimensions.

3. Pick up the tape or string. Try the same area investigation in a different part of the room. Once again, estimate, mark, measure, draw, and compare.

4. Now estimate to mark off a 12-square-foot rectangle with a *different* shape. As before, estimate, mark, measure, draw, and compare.

Think about your area estimates. Describe the strategies you used. How did you use what you learned in your first estimate to make your later estimates? Explain. Talk about this investigation with your classmates. Challenge each other to mark off rectangles with other areas.

❓ □ What It's Worth

Connections: *money; visual/spatial reasoning; measurement; functions*

Teaching Tips:

◆ Obtain a roll of quarters from a bank to bring to class. Let students feel the weight of the roll (which contains $10, or 40 quarters) to provide a benchmark they can use to help them make their estimates.

◆ Students will readily understand how unfeasible it is for them to verify their estimates by actually counting the number of quarters equal to their weight. Guide them to suggest alternatives that *are* possible to do in class. For instance, if you have access to a scale calibrated by ounces (such as a postage scale), students can use it to determine the weight of a small number of quarters and extrapolate from there. *(5 quarters ≈ 1 ounce)*

◆ A stack of 20 dimes ≈ 1 inch (question 5).

✔ *Estimates will vary.*

□ Intersections

Connections: *strategy formulation; statistics; functions*

Teaching Tips:

◆ Discuss factors that affect traffic flow, such as times of rush hour, accidents, inclement weather, length of traffic signal cycles, availability of alternative routes, etc. Brainstorm ways to carry out the investigation, such as sampling the number of cars that pass through during short periods of time and extrapolating from there.

◆ The main purpose of this investigation is to have students explore and fully develop a strategy they might use to complete such an estimate. It is not necessary to collect the data to verify the estimate, although some students may be able to do so.

◆ If possible, invite an official from the department of traffic or transportation in your area to speak to your class about how such estimates are made by professionals, and how and when these estimates are used.

✔ *Strategies will vary.*

Name_____

What It's Worth

What's the value of a bag of quarters that weighs what you weigh?

1. First, make a ballpark guess. Write it down.

2. Next, share estimates with the class. Discuss strategies you could use to improve your guess. Try one or more of them. Record your revised estimate.

3. Now, figure out a way to weigh enough quarters to check your estimate. Use any tools you find helpful, such as a scale or a calculator, as you figure the value of your weight in quarters.

4. Compare your first estimate with the value you calculated. What did you find?

5. Conclude the investigation by using the strategies you have developed to answer this question: *What is the value of a stack of dimes that's as tall as you are?*

Name_____

Intersections

What's the busiest intersection in your neighborhood? How many cars would you estimate pass through during an hour of its busiest time? How many cars do you think pass through in a day?

1. Talk about this with your group. Pick an intersection. Discuss factors that affect the amount of traffic. When do you think your intersection has the most traffic? Why?

2. Guess how many cars use the intersection during its busiest time. Guess how many use it in a day. Record your guesses.

3. Plan strategies you could use to improve your guesses. Share your guesses and strategies with classmates.

4. Talk about ways to use your plans and ways to find out actual data about traffic flow.

☐ Building Height

Connections: *visual/spatial reasoning; measurement*

Teaching Tips:

◆ If there are no tall buildings visible from your school, consider taking a field trip to where they are. Another alternative is to have students estimate the height of a flagpole, radio tower, or other structure, even a tree.

◆ List the various benchmarks students suggest, and the approximate height of each. Ask students how they know the height of these benchmarks, and when and how they may have used them.

◆ To help students find the actual heights of buildings in question, guide them to contact the building owner, superintendent or someone on the maintenance staff, or a tenant.

✔ *Estimates will vary.*

☐ Waterways

Connections: *measurement; data collection; statistics*

Teaching Tips:

◆ If necessary, review the units of liquid measure (ounce, cup, pint, quart, gallon; or liter). If you have a sink in your classroom, provide measuring cups so that students can make benchmark estimates to help them work through the questions.

◆ To answer problem 4, students will need to estimate class averages. Review mean, media, and mode, as necessary. Have students discuss the various strategies they used to make this estimate.

◆ To extend this investigation, brainstorm with the class as many other uses for water, such as bathing, laundry, cooking, agriculture, recreation (filling a swimming pool), car washing, etc. Talk about ways to estimate how much water is used in each case.

◆ You can make connections between this investigation and ecology/science issues, such as water conservation, or social studies/civics issues, such as how water is delivered to homes in your area and how municipal water supplies are kept safe for drinking.

✔ *Estimates will vary.*

Building Height

How high (in feet) is the tallest building or structure you can see from your school?

1. Name the building or structure. Make a guess and record it.

2. Discuss ways to make a closer estimate by using heights you already know. How about the height of your classroom ceiling? What about a basketball hoop, a window, or the height of another building? Which are the most useful to you? Why?

3. Now choose a benchmark you think you can use effectively. Approach the building or structure and use your benchmark to revise your estimate of its height. Record the new estimate.

4. Find out the actual height of the building or structure. How can you do this? How close was your estimate?

Waterways

Thirsty? Take a drink of water. How much water will you drink? Estimate, then talk about it.

Work together to estimate answers to the watery questions below. Think about what you know and what you need to find out. It may help to have a calculator and measuring cups at hand.

1. How much water do you drink in a day? In a month? In a year?

2. How long does it take you to drink enough to fill a bathtub?

3. How long would it take you to drink enough to fill a swimming pool?

4. How long would it take the class to drink enough water to fill a bathtub? A swimming pool?

☐ Submarine Sandwich

Connections: strategy formulation; functions; visual/spatial reasoning; measurement

Teaching Tips:

◆ The submarine sandwich has other names in different regions of the country (hoagie, grinder, hero, etc.). Be sure students understand that this investigation refers to the large sandwiches customarily served on long rolls. However, feel free to adjust the investigation to reflect foods your students may be more familiar with—tacos, bagels, burritos, egg rolls, etc.

◆ This investigation, a Fermi problem, may seem overwhelming to deal with at first, yet it is possible, once the task has been broken down into smaller steps. Students will need to do some research in order to make reasonable estimates. Questions that can guide them through this process are given on the student page.

◆ Provide a road map or atlas. Some students may need help using the map scale. They might be more successful referring to the mileage chart that often appears in a road atlas.

◆ There are many steps in this investigation. Help groups find ways to divide the tasks so that every member is actively involved and can contribute particular information that helps the group make decisions in order to estimate solutions to the problems. For instance, some students might focus on finding the distance between school and Washington, DC. Others might concentrate on calculating food estimates, and so on.

◆ You might want to assign this investigation for homework, or begin it one day, allow time for students to gather data for homework, then complete it the next day.

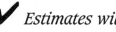 *Estimates will vary.*

Name_____

Submarine Sandwich

How many submarine sandwiches would be in a line that stretches from your school to the White House in Washington, D.C.?

- ◆ How many slices of cheese would be used?
- ◆ How much meat?
- ◆ How many tomatoes?
- ◆ How much lettuce?
- ◆ How much mayonnaise?
- ◆ What would be the total cost of this sandwich path?

This investigation may appear too hard to solve at first. But if you break it down into small bites and gather a few essential slices of information, you'll be able to make a reasonable estimate.

- ◆ You'll need to find out how far it is to our nation's capital. How will you do this?
- ◆ You'll need to estimate the length of a typical submarine sandwich, and how much of each kind of food is on one.
- ◆ You'll need to estimate about how much a sub costs.

Use any tools, strategies, or calculation methods you want. Record information as you go along so you can share your ideas later.

NOTE: You might want to work on this investigation *after* lunch.

☐ Moon Walk

Connections: strategy formulation; time; measurement; proportional reasoning; functions

Teaching Tips:

◆ This Fermi problem requires, among other things, a suspension of disbelief. Help students focus on the whimsy without getting bogged down in the scientific impossibility of the moon walk.

◆ The mean distance (to account for its elliptical orbit) from Earth to the moon is 238,857 miles. If necessary, review the relationships among measures of distance and among units of time.

◆ Students may enjoy debating whether moon walk estimates should take into account time to rest, sleep, eat space food, star-gaze, etc., or whether they should imagine being perpetual walking machines. Provide class time to discuss and come to consensus on these issues so that the estimates will be comparable. Or, you might allow groups to choose their own parameters, which they can describe when they share their estimates.

◆ To help students determine their walking pace, they might walk around a measured track at your local high school, or walk off a measured distance around your school. Talk about how fast they should walk when they determine a walking pace—will they stroll, speed-walk, or use a moderate pace?

◆ Help groups organize themselves so that they can determine walking rates. For example, two students might be responsible for laying out or measuring off a walking course. Another student can be the timer, another the walker. Students can rotate tasks to find an average walking rate for the group, which they can use to make the moon walk estimate.

◆ A reasonable range for jogging speeds would be one mile in 9–12 minutes.

✔ *Estimates will vary; at a rate of 3 mph, assuming constant walking, it would take about 9 years; a jogger could do it in 4 1/2 years.*

Moon Walk

If you could walk to the moon, about how long would it take?

Huh?

Here's another investigation that, at first, may seem impossible to do. But, again, if you take it apart, step by step, you'll be surprised at how quickly you'll be off and running.

You really only need to figure out two pieces of information: how fast you walk and how far it is to the moon.

◆ The distance to the moon is easy to find. What will you do?

◆ How can you determine your walking speed? What tools will you need?

◆ How can you use what you learn about walking speed and distance to the moon to estimate the time it would take you to get there?

Compare your estimate with those of your classmates. What factors might cause any differences you find? Talk about it.

Then get in shape so you can work together to figure out an answer to this question: *How long would it take you to jog to the moon?*

HINT: How long would it take you to jog one of those benchmark distances you used to estimate your walking speed?

Day at the Theme Park

Connections: measurement; data collection; proportional reasoning; money; map reading

◆ This investigation involves situational problem solving, much like the real-life estimation adults do to plan family trips or teachers do to plan class trips. In this case, the students are asked to take on the task. It is likely to take more than one class period, so you may want to assign parts of the investigation for homework. Or you can break it into parts that different groups can investigate separately, then pool their information to make their final plan.

◆ Introduce this investigation by having students identify nearby theme parks, living history museums, or other destinations for class trips they would enjoy. If no such parks exist in your area, you might "move" your school to another location that would be within a reasonable distance from a well-known park elsewhere. Discuss these options with the class.

◆ Collect brochures or ads from the paper to help students estimate costs. Provide road maps students can use to help them plan their routes. They may also need to gather data about gas mileage, bus rental costs, public transportation fares, etc.

◆ There are many steps involved in completing this investigation. Encourage groups to divide the tasks so that everyone is actively involved. For example, once the group has decided where to go and how they'll get there, several students might focus on the questions in part 1, while others work through the problems of part 2. One student can be in charge of writing the letter to parents, another can prepare the budget that accompanies the letter, and a third can make the final presentation to the class.

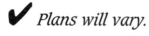 *Plans will vary.*

Name_____

Day at the Theme Park

Use estimation to plan a class day trip to a theme park.

Consider these questions as you plan for transportation and costs:

1. Getting there and back

◆ Where is the theme park? How far away is this?

◆ How will you transport the whole class?

◆ What route will you take?

◆ About how long might the trip take, one way?

◆ What time should you leave school to get there at a reasonable time? Why?

◆ When should you leave the park to return to school? Why?

2. Expenses

◆ About how much will transportation cost? Will you need to buy gas? Are there tolls?

◆ What is the park entrance fee? Is there a discount for a large group like yours? What about parking fees?

◆ Do rides cost extra? If so, about how much? About how many rides will each student go on?

◆ How much should you budget for food and drinks? What about souvenirs or anything else students may wish to buy?

Write a letter to parents outlining your plan for this trip. Include a schedule and all estimated costs. Tell how much money each student should contribute to group costs, as well as a suggested amount per student to cover other expenses. Share your plan with the class. Explain your decisions.

Classroom Movers

Connections: visual/spatial reasoning; time; geometry; measurement; proportional reasoning

Teaching Tips:

◆ Introduce this investigation by talking about moving. Any students who have moved can share their experiences with the class. If possible, display some empty cartons to help students visualize about how much will fit and estimate how many they will need.

◆ Students may need help deciding where to begin. Brainstorm ideas with the class that groups could use to get started.

◆ Some students may actually pack a box with books or other typical classroom materials to create a hands-on benchmark they can use to refine their estimate. Remind them that they will have to carry the boxes they pack, so they should not overload them.

◆ Many large moving companies provide brochures or pamphlets with tips for packing and moving. Contact them to see if you can obtain some of these for students to use to help them in this investigation. If possible, invite a professional mover to visit the class and make the same kind of estimate the students did. Encourage your visiting specialist to share any techniques or benchmarks customarily used to make accurate estimates. After, allow students to rework their estimates, taking into account what they learned.

 *Estimates will vary.*

Grapefruit Volume

Connections: visual/spatial reasoning; measurement; geometry; proportional reasoning

Teaching Tips:

◆ This investigation, while whimsical, involves serious estimation strategies. Students must consider spherical objects that do not fill a space completely.

◆ Make sure students understand the premise of this investigation—the room will be completely vacant of everything that can be moved. The citrus-loving principal will fill the room with grapefruits right up to the ceiling, as if it were one giant crate.

◆ Bring in several grapefruits to display in class. Provide rulers, calculators, and any other tools students might want to use to help them make their estimates. One strategy students might use is to fill a box with grapefruits, estimate how many same-size boxes would fit in the room, and adjust the estimate.

 *Estimates will vary.*

Name _____

Classroom Movers

Good news! You're getting a brand new classroom. It's bigger and better, and it's right down the hall.

BUT—you have to move from the classroom you're in now. You don't have to move furniture, but you *do* have to move everything in it, on it, or near it. Don't forget things stored in closets, drawers, and cabinets.

Make estimates about your move, just like professional movers do.
- ◆ About how many boxes will you need?
- ◆ About how long will it take to pack the boxes?
- ◆ About how long will it take to move the boxes?

Write a description of your plan for the move. Describe how you would use benchmarks to answer the questions. Post your plan for classmates to examine.

Name _____

Grapefruit Volume

Your principal has an unusual plan for the room your class is vacating—she's going to pack it full of grapefruits.

- ◆ About how many grapefruits would it take to fill your room?
- ◆ How will you figure it out?

But what if you could convince her to fill it with something else—for example, bowling balls, tennis balls, or beach balls?

- ◆ How many of each would you estimate could fill the room?
- ◆ How can you use what you know about grapefruits to figure it out?

☐ Brick Wall

Connections: visual/spatial reasoning; measurement; functions

Teaching Tips:

◆ Review measurement relationships (inches, feet, yards), as needed.

◆ A typical brick is 8 inches long, 2 3/4 inches high, and 3 inches deep. Provide this information, or have students measure bricks in their neighborhood. If possible, display bricks so students can use them as visual references.

◆ Point out that students must account for the mortar between bricks that holds them together. As a rule of thumb, allow 1/2 inch of mortar between bricks. Students should consider how to treat the corners of the wall. Encourage any reasonable adjustments.

◆ A football field is 100 yards long (plus 10 yards for each end zone) by 53 yards wide. Discuss these dimensions with the class. You should also come to consensus about how many bricks deep the wall will be so that all estimates are based on comparable data.

◆ Some students may use the "solve a simpler problem" strategy: design a small section of wall, calculate the number of bricks it takes, and extrapolate from there.

✔ *Based on a rectangle 120 ft x 53 ft, 22,000 bricks is a reasonable estimate for a wall one brick deep.*

☐ Bicycle Census

Connections: strategy formulation; number sense; logical reasoning; data collection; statistics

Teaching Tips:

◆ This Fermi problem requires students to think about who uses bicycles, how many bicycles there might be per household, and the different kinds of businesses that use them (i.e., messenger/delivery services, cycle police, rental companies, parks, etc.).

◆ Students can use almanacs or other resources to get information about the number of households in your state. Discuss ways to estimate the number of businesses that use bicycles, and how many each might use. You might want to assign the research for homework, or have students work on this investigation with family members.

◆ Guide students to adjust their estimates as they see fit to account for households without bicycles, those with more than one bicycle, or businesses that use many bicycles. Feel free to substitute other things to estimate, such as number of pets, TV sets, refrigerators, or cars in your community, county, or state.

✔ *Estimates will vary.*

Name _____

Brick Wall

Imagine that your local planning board has decided to build a low brick wall around a neighborhood park. It is taking bids from local construction companies. The company that most accurately estimates the number of bricks needed gets to build the wall.

◆ The park is a rectangle the same size as a football field.

◆ The wall will be about 4 feet high.

◆ There must be a 3-foot-wide entrance in each side of the wall.

Give your construction group a name. Then develop and use an estimation strategy to predict the number of bricks needed. Submit your estimate and a written description of your strategy.

Whose group gets the contract to build the wall?

Name _____

Bicycle Census

Not counting bicycles in warehouses, factories, or on display in stores, how many bicycles are there in your state?

Use your number sense and your common sense to make a smart guess. Who uses bicycles? Brainstorm a list of the kinds of people who use bicycles or the places where you might find bicycles in use. Remember, it's not just kids who ride bikes.

◆ Try to break this investigation down into a series of smaller problems. How might you do this? How will it help to do this?

◆ What data about your state would be useful for you to have? How can you get the information you need?

Compare your estimates with those of other groups. How can you find out which estimate is the closest to the actual figure?

☐ **Estimation Celebration** *(student pages on 71-72)*

Connections: *visual/spatial reasoning; time; measurement; number sense; geometry*

Teaching Tips:

◆ The purpose of this final investigation is to wrap up the exploration of estimation with a fun-filled array of estimation events. You might want to set aside time to plan the events. It may require a double session of class time to hold the celebration.

◆ Before the celebration, talk with the class about parameters everyone can agree on. For example, how will estimates be judged, measured, or recorded? Where will events take place? How can the room be arranged to optimize the space available for the events? How many times can each student try each estimate? Who will gather the necessary materials? List the questions that students pose, and come to a consensus before holding the celebration.

◆ Students can work in groups or as individuals. They can try every event or choose several. Or, you can divide the class into teams that "compete" in the events, in the spirit of a class Olympics.

◆ Encourage students to suggest other events to add to the celebration, or replace some of the ones here with others that may be more suitable to their interests and backgrounds. For example, if your students don't know the songs suggested in event 8, have them suggest one multiple verse song they *do* know.

◆ An estimate celebration can make for another interesting possibility for a math fair or family math night.

◆ Have fun!

 *Estimates will vary.*

Name _____

Estimation Celebration

Now that you've become expert estimators, it's time to celebrate your new skills. Have an estimation party, but instead of playing the usual party games, try some of these activities. Don't worry, you'll never find questions like these on a test!

Estimate first. Record your estimate, then check by trying the activity. Who's the champion estimator in your class?

1. How long can you balance on one foot, with the other foot bent at the knee, thigh parallel to the floor?

2. How long can you balance on tiptoe with your arms outstretched above your head?

3. How high can you build a stack of pennies?

4. How close to 5 feet can you flick a penny along the floor?

5. How far can you jump forward from a standing position?

6. How far up a wall can you reach?

7. How long can you balance a ruler vertically on your palm?

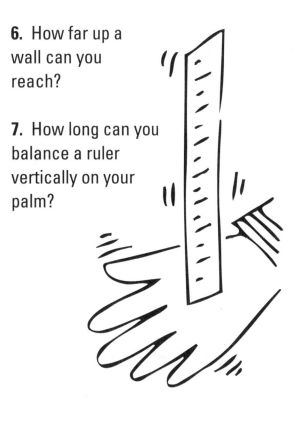

8. How long will it take you to sing all the verses of "Found a Peanut" or "100 Bottles of Pop on the Wall"?

Estimation Celebration (continued)

9. How long is your wingspan?

10. How many sheets of paper are in your notebook or loose-leaf?

11. How far away can you stand from a dictionary and still read an entry word and its definition?

12. How long will it take you to make a paper-clip chain as long as you are tall?

13. Altogether, how many buttons are you and your classmates wearing today?

14. How many pencils rubber-banded together have a diameter of about 2 inches?

15. How many cotton balls can you stuff into a paper cup without any of them popping out?

16. How long a continuous strip can you make by cutting up a 3 x 5 index card? (NO TAPE ALLOWED!)

PART 4

Appendix

Computational Estimation Strategies

USING THE STRATEGIES

When students compute with numbers, they can either find exact answers or make estimates. The following chart illustrates the distinction between the two approaches. The difference is that to estimate, problem solvers alter the numbers of the problem in some way to make them more manageable. Keep in mind that whether problem solvers want to find exact answers or estimates, they'll still use one or more of the three basic computation methods—mental math, calculators, or paper and pencil. The method used to find an exact answer will yield an exact answer; the method used to find an estimate will yield an approximation.

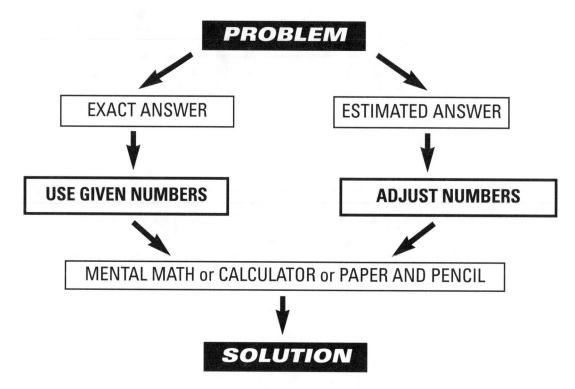

Math educators have identified several useful strategies for computational estimation. Effective estimators know that no one strategy works for every problem and that some problems may require more than one strategy. They are able to draw from a variety of strategies, and apply those that will work most efficiently in a given situation. Successful problem solvers understand that the goal when using any of these strategies is to change the given numbers in a problem so that they are made easier to work with. Or they can alter the structure of a problem to manipulate the numbers more effectively.

Changing numbers in a problem is likely to make the estimate greater or less than the exact answer would be. Skilled estimators are able to judge whether their estimate is high or low and then make an adjustment by adding to or subtracting from it.
This refining process is called *compensation*. Encourage students to use compensation along with any of the computational estimation strategies.

FRONT-END ESTIMATION

This strategy, which focuses on the left-most digits of the numbers involved, can be easily used with all four operations. It is among the most commonly used strategies, and is often the first one introduced to younger children.

Steps

1. Compute using the front-end digits only.

2. Adjust the answer.

Examples

A. Estimate the total of this grocery receipt:

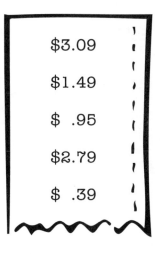

$3.09

$1.49

$ .95

$2.79

$ .39

Think: Find the front-end total. The dollars total $6. Ignore amounts less than $1. Adjust by grouping the cents to form dollars. The cents add to about $3.

Since $6 + $3 = $9, the grocery receipt totals about $9.

B. Estimate the cost of 4 cans of olive oil that sell for $4.55 each.

Think: 4 x $4 is $16, and 4 x $.55 is about $2.

So, 4 x $4.55 is about $16 + $2, or $18.

FLEXIBLE ROUNDING

Numbers can be rounded in many ways, according to the numbers involved, the situation, and the skills and number sense of the estimators.

Steps

1. Substitute manageable numbers by rounding the numbers to a given place, such as to the greatest place, the nearest dollar, or to a sensible multiple of 10.

2. Compute with the rounded numbers.

3. Adjust to compensate for underestimates or overestimates, as necessary.

Examples

A. Estimate the total: $5.95 + $6.99 + $2.39 + $3.45 + $8.25 = ?

Think: Round each amount to the nearest dollar. Add the dollars.

$6 + $7 + $2 + $3 + $8 = $26

But $26 is an *under*estimate. Adjust by flexible rounding.

$6 and $7 are close estimates to $5.95 and $6.99. But the next three amounts were underestimated by a total of about $1. (39¢ + 45¢ + 25¢ ≈ 100¢, or $1)

So, add $1 + $26, for a total estimate of about $27.

B. Estimate 2687 ÷ 92.

Think: Round 2687 to its greatest place, 3000.

Round 92 to 100. Estimate: 3000 ÷ 100 = 30

So, 2687÷ 92 is about 30.

C. Estimate 55 x 86.

Think: Round 55 *down* to 50 and 86 *up* to 90.

If one number is rounded up and the other one rounded down, the estimate has already been adjusted.

50 x 90 = 4500

So, 55 x 86 is about 4500.

COMPATIBLE NUMBERS

Using compatible numbers means replacing the given numbers with numbers that are easy to manipulate mentally. This strategy requires flexibility in rounding. It is particularly useful when dividing or when adding a column of numbers.

Steps

1. Use number sense to choose a set of compatible numbers that fit the given numbers.

2. Compute with these substituted numbers.

Examples

A. Estimate $4109 \div 6$.

Think: Find a pair of numbers that are easy to manipulate, such as $4200 \div 6$.

Use mental math to divide: $4200 \div 6 = 700$

So, $4109 \div 6$ is about 700.

B. Estimate the sum:

$$
\begin{array}{r}
41 \\
87 \\
22 \\
14 \\
79 \\
+\ 61 \\
\hline
\end{array}
$$

Think: Restructure the problem. Look for pairs of numbers that "fit" in some way. One idea is to look for pairs that make about 100.

$41 + 61$ is about 100, $87 + 14$ is about 100, and $22 + 79$ is about 100.

$100 + 100 + 100 = 300$

So, the sum is about 300.

? CLUSTERING

This strategy, also known as *averaging*, is useful when a group of numbers clusters around a common value.

Steps

1. Estimate to choose a reasonable group average.

2. Multiply that average by the number of values.

Examples

A. The 5-game baseball series had the following attendance figures:

Game	Attendance
1	49,259
2	53,208
3	51,992
4	47,996
5	53,563

Estimate the total attendance for the series.

Think: All attendance figures are *about* 50,000.

There are 5 games. 5 x 50,000 = 250,000

So, the total attendance was about 250,000.

B. This table shows monthly snowfall totals last winter.

Month	Snowfall in Inches
December	13.83
January	15.1
February	16.45
March	16.3

What was the total snowfall for the four months?

Think: All amounts cluster around 15 inches.

4 x 15 = 60

So, the total snowfall was about 60 inches, or 5 feet.

BENCHMARK ESTIMATION

This strategy, which works particularly well with fractions and percents, involves finding and using benchmark numbers to estimate answers. When working with decimals, powers of 10 make useful benchmarks.

Steps

1. Round to choose a benchmark or benchmarks.

2. Use mental math to compute.

Examples

A. Estimate 4/9 + 1 7/8.

> **Think:** 4/9 is about 1/2. 1 7/8 is about 2.
>
> Add: 1/2 + 2 = 2 1/2
>
> So, 4/9 + 1 7/8 is about 2 1/2.

B. Estimate the number of blue T-shirts sold, if 47% of the 160 shirts sold were blue.

> **Think:** 47% is about 50%.
>
> 50% of 160 is 80.
>
> So, somewhat fewer than 80 blue T-shirts were sold.

C. Estimate 1102.374 x 19.2

> **Think:** Use powers of 10.
>
> 1102.374 is about 1000.
>
> 1000 x 19.2 = 19,200
>
> So, 1102.374 x 19.2 is about 19,200.

NOTES